MORNINGSIDE

The cover picture shows the six buildings of Morningside Gardens, initially occupied June 24, 1957. In the background are buildings of the N.Y.C. Housing Authority General Grant Houses.

Initial occupancy ceremony.

In foreground Borough President Hulan E. Jack and David Rockefeller.

Morningside Gardens

JULY 15, 1957

REPORT ON NEW YORK CITY SLUM CLEARANCE PROGRAM UNDER TITLE I OF THE FEDERAL HOUSING ACT OF 1949

Until 1949, when New York City began its program for large-scale clearance and redevelopment of slum areas, practically all public slum clearance and rebuilding of blighted areas in the City was done by the New York City Housing Authority in connection with low-rent liberally subsidized public housing. The problem of attracting private capital to the risky field of slum clearance and rebuilding of the blighted or deteriorating areas was attacked under the provisions of the Title I of the Federal Housing Act of 1949.

This act provides Federal aid to a City or other local public agency in acquiring slum sites and redevelopment by private capital pursuant to plans approved by the local authorities and the Federal Government. Aid is in two directions—first, the expansion of the power of eminent domain, that is condemnation to acquire slum property in many ownerships which could not be assembled by ordinary purchasing procedure, and second, writing down the price of land to be acquired at a figure private capital can afford to pay. In a rough way this write down represents the cost of buildings which must be torn down and have no value to developers. The write down losses incurred by a City or other local agency as a result of the acquisition and resale of sites are shared, two-thirds by the Federal Government and one-third by the locality.

The Committee on Slum Clearance was appointed by Mayor O'Dwyer on December 17, 1948 to study and expedite the promotion of specific slum clearance projects under the anticipated Federal law later enacted as Title I of the Housing Act of 1949. A preliminary report was made on July 4, 1949 and the Committee was instructed to continue its studies and prepare a definite program for public discussion. On January 23, 1950 a further interim report was made. Reports on actual progress on the program adopted were made in November 1953 and April 1956.

The Federal funds authorized under the 1949 law were limited to $500,000,000 for capital grants throughout the entire nation, at the rate of $100,000,000 a year for five years ending in 1953. No State could receive more than $50,000,000 or 10% of the total available. New York City was allocated some $40,000,000 of New York State's $50,000,000 maximum. The enactment of Public Law #94 in 1953, at our request, made available to the City an additional $25,000,000 out of the original $500,000,000 making a total of $65,000,000. This meant that with the City's one-third contribution, about $97,500,000 was available to absorb losses. In view of the areas in the City requiring slum clearance, once estimated at as much as 9,000 acres, these funds were not sufficient to do more than blaze the way and to finance pilot projects as examples for a large future program.

The Housing Act of 1955 provided an additional $500,000,000 in Title I capital grant authorizations for the years 1955 through 1957 throughout the nation. This included provisions for $100,000,000 to be used at the discretion of the President which, however, he has not as yet made available. New York City has received to date about $39,000,000 in additional capital grant reservations under the 1955 Act. The Committee has recently conferred with the Administrator of the Housing and Home Finance Agency and a general understanding has been reached under which further allocations will be made to the City to provide for the excess requirements of our present program and permit the preparation of plans for additional projects. Under the program agreed upon Federal write down funds heretofore allocated and reserved under the 1949 and 1955 Housing Acts, will be increased to a total of $110,000,000. In addition about 10% and perhaps more of the amount of capital grant authorized by Congress in 1957 will be made available to New York City after July 1, 1957. We have already completed planning and are in the process of preparing studies of a program of Title I slum clearance projects in the City which more than exhausts this sum.

The Committee on Slum Clearance has to date submitted to the Board of Estimate reports on nineteen Title I projects. Of these, ten are operating projects in which the land has already been acquired and resold to redevelopers who are in possession and carrying out the redevelopment plans. These ten projects are: Corlears Hook, Harlem, North Harlem, West Park, Morningside-Manhattanville, Columbus Circle, Fort Greene, Pratt Institute, N.Y.U.-Bellevue and Washington Square Southeast.

CORLEARS HOOK

ILGWU Cooperative Village-Corlears Hook Title I project, four buildings at center of picture. East River to left, Williamsburg Bridge across picture.

Initial occupancy ceremony.

David Dubinsky speaking.

They will provide over 15,000 new dwelling units, will clear approximately 162 acres of land at a written down cost of about $84,000,000 of which $56,000,000 will be the Federal and $28,000,000 the City share. Private builders have contracted to pay about $26,000,000 for the land, and will spend an estimated $245,000,000 in construction and development costs. Taxes before acquisition were computed to be about $2,245,000 and it is estimated that after completion they will be approximately $5,778,000 at the current rates.

Reports on the following five projects in the advance planning stage are being processed by the Housing and Home Finance Agency, and approval of capital grant contracts are expected shortly. They are: Lincoln Square, Seward Park, Park Row, Seaside and Hammels. These projects will provide over 10,000 new dwelling units, will clear approximately 150 acres at a written down cost of about $66,000,000 of which $44,000,000 will be shared by the Federal and $22,000,000 the City's share. The Committee has sponsors for all these projects and it is estimated that about $232,000,000 will be spent in construction and development costs. Present tax returns in these areas amount to $1,566,000 and the new return will amount to about $4,615,000.

Of the four remaining projects reported on. Washington Square South and South Village were filed by the Board of Estimate. The Williamsburg project, which did not attract a redeveloper is no longer under consideration. The Delancey Street project, which was originally proposed in 1951 and recently restudied to make desirable revisions in the redevelopment plan has been deferred because the Federal Agencies do not regard the area as fully acceptable for rental housing.

Six new projects in the preliminary planning stage for which reports are now being prepared will involve a write down of about $45,000,000 of which the Federal Government will share $30,000,000 and the City $15,000,000.

These new projects are: Penn Station South, Riverside-Amsterdam, Battery Park, Gramercy Park, Soundview and Jamaica Race Track. These new projects will clear about 295 acres of land and provide approximately 12,500 new dwelling units. The City now receives $1,193,000 in taxes in these areas and it is estimated that the new return will be $3,944,000. The Jamaica Race Track project involves an area of 178 acres which will be developed as a large scale cooperative housing project with accommodations for about 4,000 families in the middle income range. Recent discussions with the Racing Association indicate that it is likely that the present track which is considered inadequate and expendable, can be made available in the next two years.

The slum clearance program in New York City is substantially larger and further advanced than any other Title I program in the country. The twenty-one projects outlined involve the clearance of over 600 acres of slum property with total write downs of about $196,000,000 by the Federal Government and the City. In addition private developers and various institutions involved in these projects will spend about $628,000,000 in construction exclusive of land costs. The new assessments to be added to our tax rolls due to these improvements should produce an additional yearly tax return of over $9,000,000 in excess of present collections.

While in a sense only a start has been made in terms of our total slum clearance problems and large additional sums must be provided to insure continuation of the program, it should be noted that substantial progress has been made toward the ultimate goal. The projects mentioned above comprise a total investment of over $881,000,000 in Federal, City and private funds and will provide over 38,000 modern dwelling units including many needed community and educational and other institutional facilities. Projects under study and those being planned for the future are estimated to require about $68,000,000 in additional Federal funds and $34,000,000 in City funds. They involve the clearance of 225 acres of slum properties and provide over 15,000 new dwelling units.

One of the most pressing reasons for proceeding quickly with this program is the fact that, at least in large part, this program will be directed at providing for the large influx of population from other areas of the country and Puerto Rico whose housing needs must be met not only by increased substantially subsidized public housing but also by new moderate rental middle income and cooperative developments.

An outstanding feature of our program is the increased number of cooperative developments planned with the unions. In our overall program it is expected that union groups will undertake the construction of projects totaling 22,000 dwelling units at a cost of some $290,000,000. These will make new modern apartments available at moderate rentals of $19 to $25 per room on the basis of tax concessions by the City. This continued and growing interest on the part of experienced union groups in the program is most significant and encouraging.

Continued on page 14

COLUMBUS CIRCLE

New York Coliseum and Columbus Circle apartments. Central Park in left foreground.

COMPLETED

Corlears Hook	Cherry, Jackson, Lewis and Delancey Streets, F. D. Roosevelt Drive, Manhattan.

UNDER CONTRACT

Harlem	W. 132nd to 135th Streets, from 5th to Lenox Avenues, Manhattan.
North Harlem	W. 139th to 142nd Streets, from 5th to Lenox Avenues, Manhattan.
West Park (Manhattantown)	W. 97th to 100th Streets, from Central Park West to Amsterdam Avenue, Manhattan.
Morningside-Manhattanville	W. 123rd to LaSalle Streets, from Amsterdam Avenue to Broadway, Manhattan.
Columbus Circle	W. 58th to 60th Streets, from Columbus Circle to Columbus Avenue, Manhattan.
Fort Greene	DeKalb Ave. and Myrtle Ave., Ft. Greene Park to Flatbush Ave. Extension, Brooklyn.
Pratt Institute Area	Myrtle and Lafayette Avenues, Classon Avenue to Hall Street, Brooklyn.
N.Y.U.-Bellevue	E. 30th to 33rd Streets, from 1st to 2nd Avenues, Manhattan.
Washington Sq. Southeast	W. 4th to W. Houston Streets, from Mercer Street to W. Broadway, Manhattan.

ADVANCED PLANNING

Seaside	Beach 102nd to Beach 108th Streets, Shore Front Parkway to L. I. Railroad, Queens.
Lincoln Square	Columbus to Amsterdam Avenues from W. 60th to W. 65th Streets. Amsterdam Avenue to N. Y. Central R.R. from W. 66th to W. 70th Streets, Manhattan.
Seward Park	Grand Street, E. Broadway and Essex Street, Manhattan.
Park Row	Park Row, St. James Place and Pearl Street, Manhattan.
Hammels	Beach 74th to Beach 90th Sts., Rockaway Beach Blvd. to Shore Front Pkwy., Queens.

PRELIMINARY PLANNING

Penn Station South	W. 23rd to W. 29th Streets from 8th to 9th Avenues, Manhattan.
Riverside-Amsterdam	W. 83rd to W. 86th Streets from Amsterdam Avenue to Riverside Drive, Manhattan.
Battery Park	South to Water Streets from Whitehall Street to Coenties Slip, Manhattan.
Gramercy Park	E. 24th to E. 27th Streets from 2nd to 3rd Avenues, Manhattan.
Soundview	White Plains Road, O'Brien Avenue and U.S. bulkhead line on south and west, Bronx.
Jamaica Race Track	New York Boulevard, Baisley Boulevard, 172nd Street to 129th Avenue, 76th Street, Garret Street, New Street and 137th Avenue, Queens.

UNDER STUDY

Bellevue South	E. 23rd to E. 30th Streets from 1st to 2nd Avenues, Manhattan.
Cooper Square	Houston Street to E. 9th Street from 2nd Avenue to Bowery and 3rd Avenue. Delancey to Houston Streets from Chrystie Street to Bowery, Manhattan.
Mid-Harlem	W. 125th to W. 135th Streets from 8th to St. Nicholas Avenues, Manhattan.
Cadman Plaza	Fulton Street, Henry Street, Clark Street, Brooklyn.

FUTURE

Seward Park Extension	Manhattan
Washington Market	Manhattan
Park Row Extension	Manhattan
Division Street	Manhattan
Arverne Rockaway	Queens
Cathedral Parkway	Manhattan
Mott Haven	Bronx
Northern Boulevard	Queens
South Brooklyn	Brooklyn
Atlantic Avenue	Brooklyn

NORTH HARLEM
On Lenox Ave. from 139th St. at right to 142nd St. at left. Harlem River in background.

WASHINGTON SQUARE S.E.

Washington Square Park at left, Houston St. at right. Below, West Broadway, above, Mercer St.

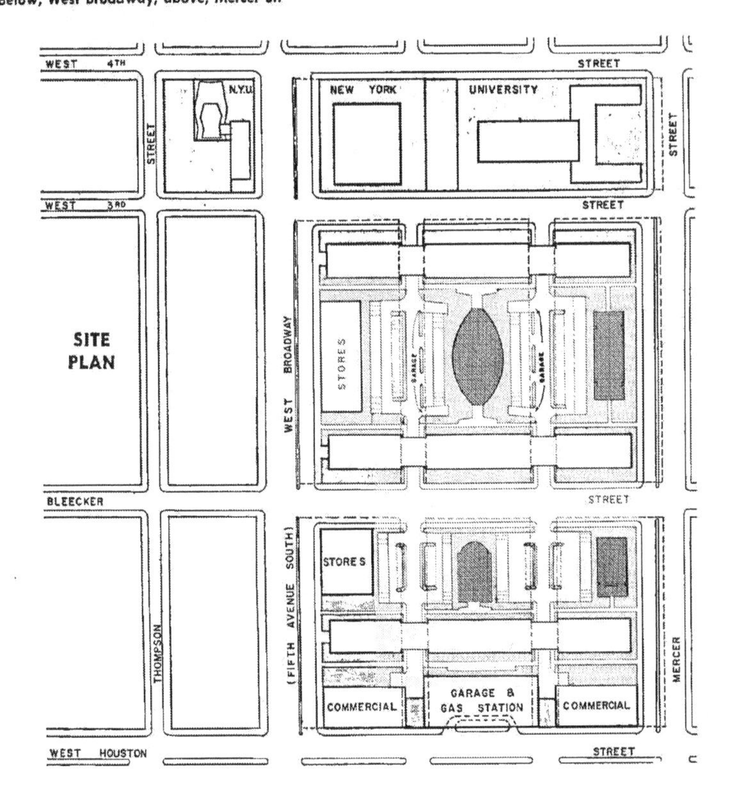

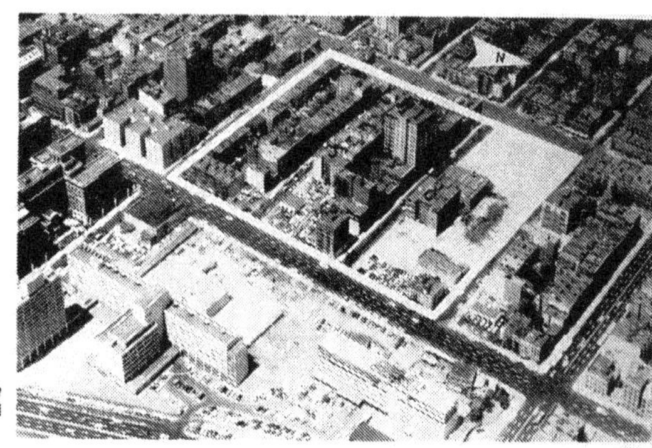

HARLEM

Fifth Ave., below, to Lenox Ave., 132nd St. at left to 135th St. Abraham Lincoln Houses and Riverton are in foreground.

←

Westerly Section of project.

WEST PARK-MANHATTANTOWN

From 100th St. at left to 97th St., Central Park to Amsterdam Ave. Public School #105 under construction at lower right. N.Y.C. Housing Authority Douglass Houses at left.

N.Y.U. BELLEVUE

First Ave., below, to Second Ave., 30th St., left, to 33rd St. East River Drive and N.Y.U. Bellevue Medical Center in foreground.

PRATT INSTITUTE

From Myrtle Ave. in foreground, to Lafayette Ave. at rear. At left is Classon Ave., at right, Hall St. Above and below will be housing with Pratt Institute occupying center section.

Ground breaking ceremony of Housing Development.

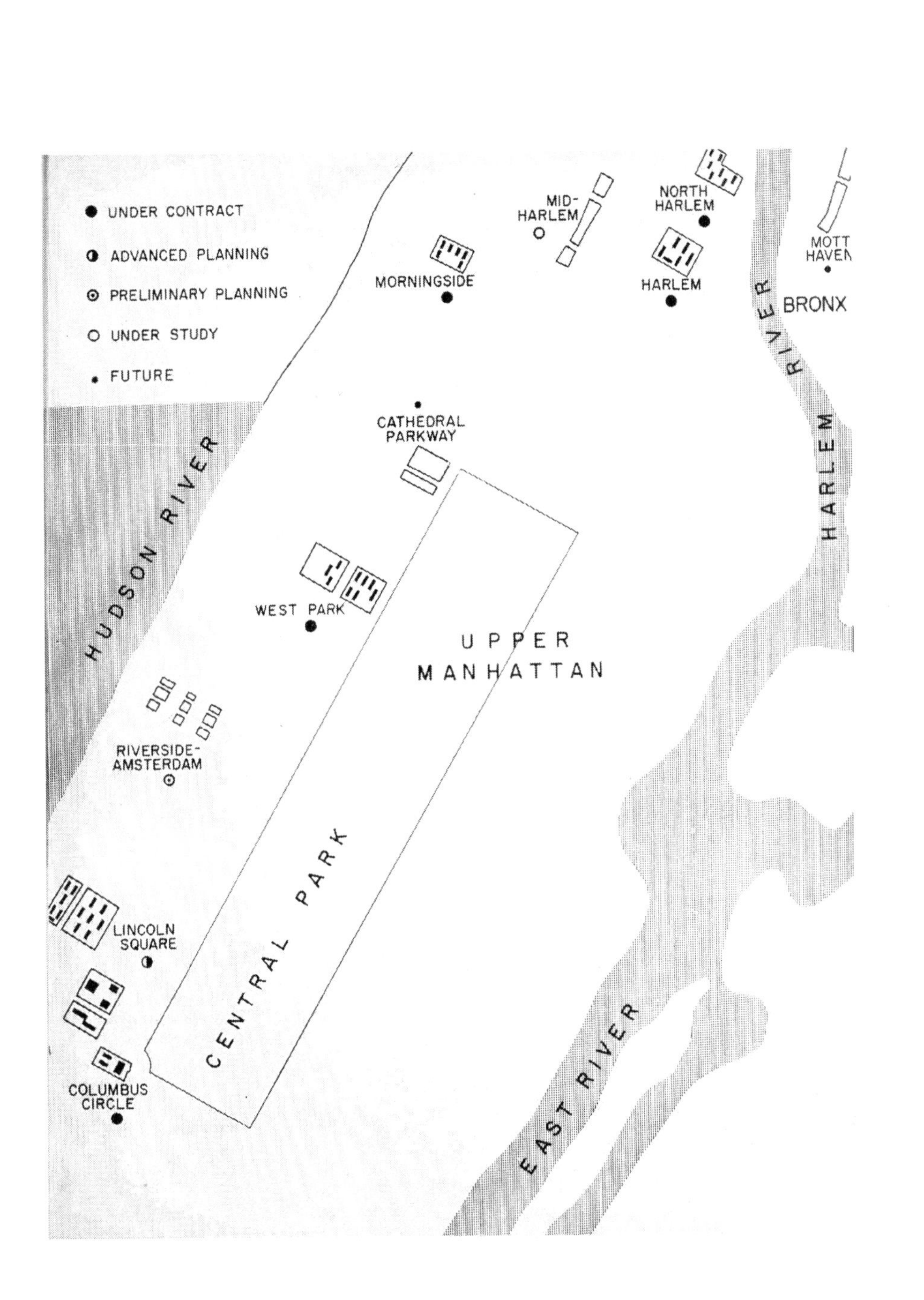
UNDER CONTRACT
ADVANCED PLANNING
PRELIMINARY PLANNING
UNDER STUDY
FUTURE
HUDSON RIVER
MORNINGSIDE
MID-HARLEM
NORTH HARLEM
HARLEM
MOTT HAVEN
BRONX
HARLEM RIVER
CATHEDRAL PARKWAY
WEST PARK
UPPER MANHATTAN
RIVERSIDE-AMSTERDAM
CENTRAL PARK
LINCOLN SQUARE
COLUMBUS CIRCLE
EAST RIVER

FORT GREENE

Ft. Greene Park to left, Flatbush Ave. Ext. at right, Myrtle Ave. in foreground and DeKalb Ave. at rear. Brooklyn Hospital is just outside white line at upper left. The five buildings of Kingsview Homes at extreme left are occupied. Foundation for University Towers Housing under way in front center.

Kingsview Homes

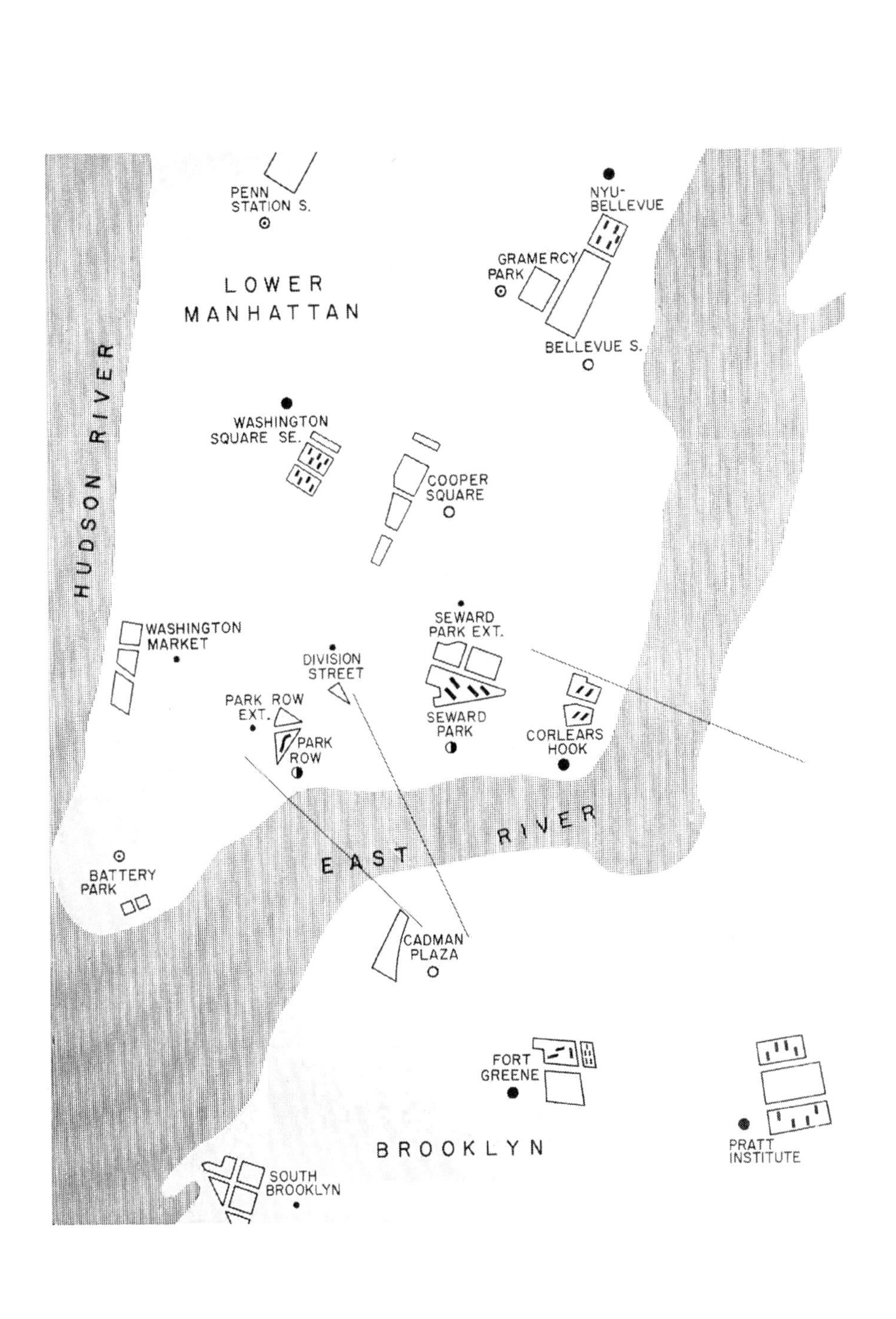
PENN
STATION S.
NYU-
BELLEVUE
GRAMERCY
PARK
LOWER
MANHATTAN
BELLEVUE S.
HUDSON RIVER
WASHINGTON
SQUARE SE.
COOPER
SQUARE
WASHINGTON
MARKET
SEWARD
PARK EXT.
DIVISION
STREET
PARK ROW
EXT.
PARK
ROW
SEWARD
PARK
CORLEARS
HOOK
EAST RIVER
BATTERY
PARK
CADMAN
PLAZA
FORT
GREENE
PRATT
INSTITUTE
BROOKLYN
SOUTH
BROOKLYN

Following is a resume of progress on the first ten sites undertaken by the Committee. The land required for these projects was acquired by the City with the approval of the Federal Government and resold to responsible builders, who are now engaged in relocation of tenants, demolition of buildings and construction of the new developments. All litigation attacking slum clearance projects has now been settled. The entire New York City program has been held legal and constitutional by State and Federal courts.

OPERATING PROJECT STATUS

Corlears Hook (ILGWU Cooperative Village)

Project completed and 100% occupied, initial occupancy October 1955. Includes shopping center and playground.

Columbus Circle

Coliseum Area—Tenant relocation completed June 1954, site cleared November 1954, construction completed and opened on April 25, 1956.

Housing Area—Tenant relocation completed, site cleared, construction 80% completed, initial occupancy expected September 1957.

Harlem (Godfrey Nurse Houses)

Tenant relocation 80% completed, site 65% cleared, one shopping section completed and construction started on first three residential buildings. Project includes shopping and new school and playground shared with North Harlem project.

North Harlem (Delano Village)

Tenant relocation 88% completed, site 55% cleared, construction on first three residential buildings 70% completed. Project includes shopping area.

West Park (Manhattantown)

Tenant relocation 73% completed, site 60% cleared. Construction to be started in July with first units occupied by October 1, 1958. Sixty-eight percent of stock of the Manhattantown Corp. has been purchased by Webb & Knapp, Inc. Project includes shopping, new school, playground, Health Center and Public Library. New school under construction.

Morningside Manhattanville (Morningside Gardens)

Tenant relocation completed, site cleared, construction 92% completed, initial occupancy June 24, 1957. Project includes shopping area.

Fort Greene

a. Long Island University Area—Tenant relocation completed, site cleared, construction not started.
b. Brooklyn Hospital Addition Area—Tenant relocation completed, site 50% cleared, construction not started.
c. Kingsview Homes—Project completed 100% occupied, initial occupancy November 1956.
d. University Towers—Tenant relocation completed, site 97% cleared, construction started. Project includes shopping and playgrounds.

Pratt Institute

a. Educational Area—Tenant relocation completed, site 93% cleared, construction of new athletic field 20% completed. Two dormitories completed in 1956. Construction started on Union Building.
b. Housing Area—Tenant relocation 82% completed, site 45% cleared, one shopping section completed and construction started on first residential building. Project includes new public school, playgrounds and shopping.

New York University-Bellevue

Tenant relocation 60% completed, site 30% cleared, construction not started. Project includes shopping area and a professional building.

Washington Square Southeast

a. Educational Area—Tenant relocation 60% completed, site 70% cleared. Provides for expansion of New York University.
b. Housing Area—Tenant relocation 41% completed, site 27% cleared, construction not started. Project includes shopping and garage.

Continued on page 18

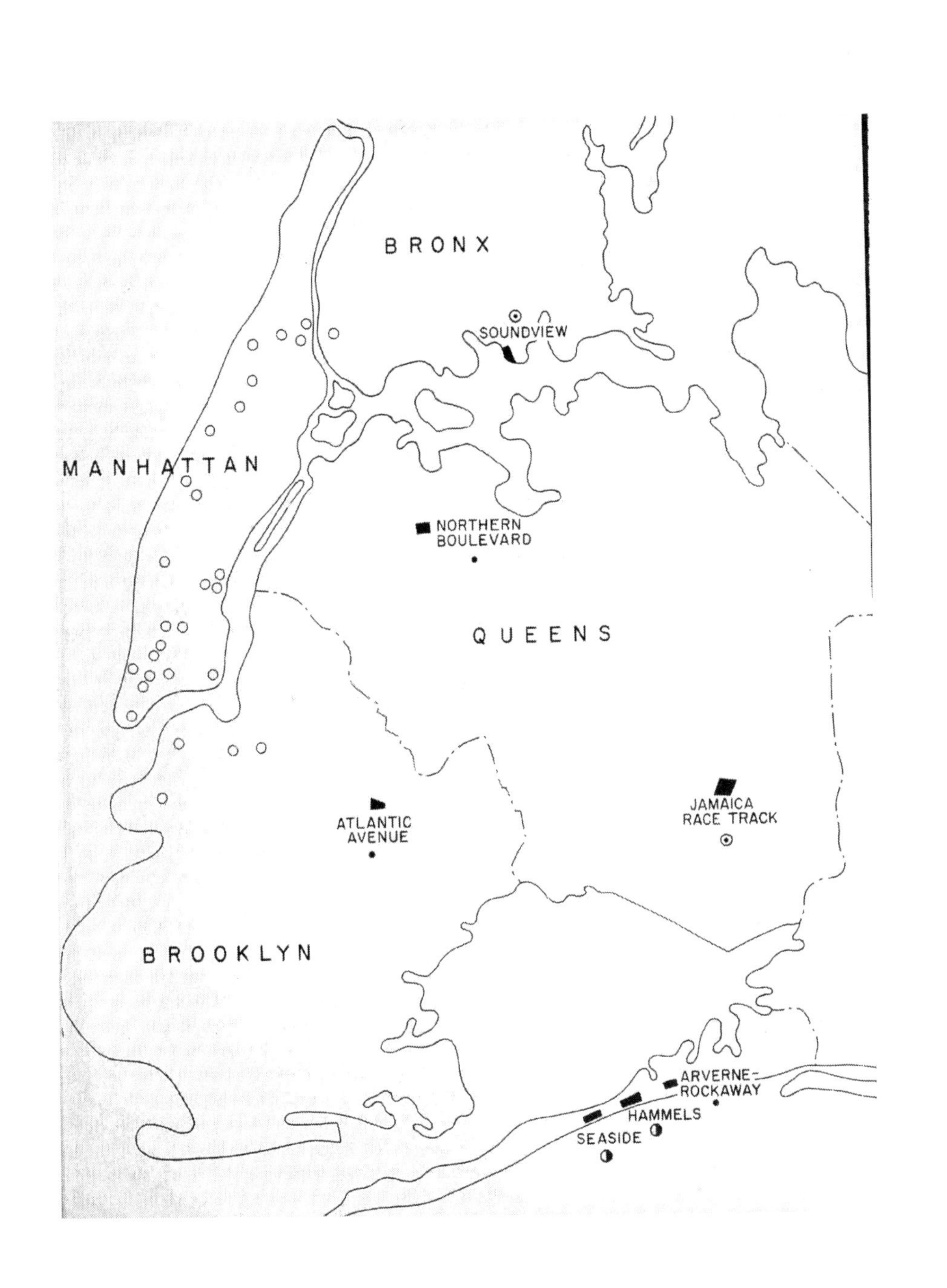
BRONX
SOUNDVIEW
MANHATTAN
NORTHERN BOULEVARD
QUEENS
ATLANTIC AVENUE
JAMAICA RACE TRACK
BROOKLYN
ARVERNE-ROCKAWAY
HAMMELS
SEASIDE

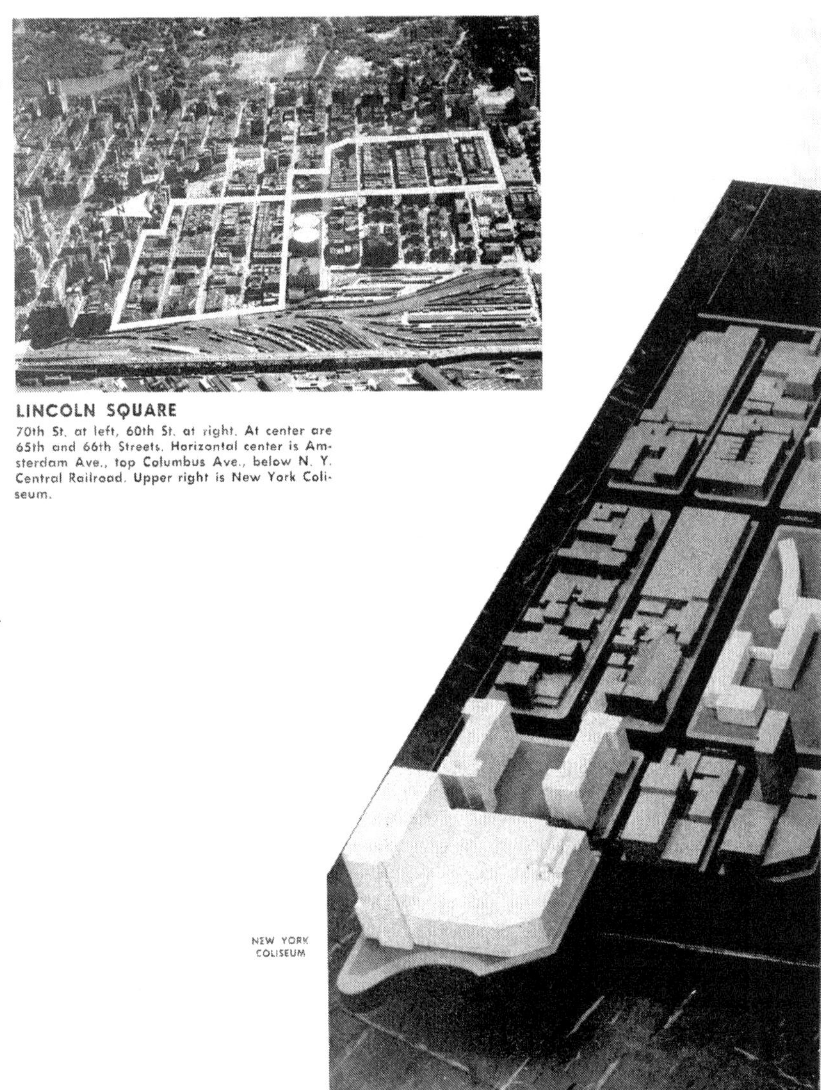

LINCOLN SQUARE

70th St. at left, 60th St. at right. At center are 65th and 66th Streets. Horizontal center is Amsterdam Ave., top Columbus Ave., below N. Y. Central Railroad. Upper right is New York Coliseum.

PERFORMING ARTS CENTER
CENTER BELOW
COLLEGIATE CENTER
TO LEFT
HOUSING

TENANT RELOCATION

The Bureau of Real Estate of the Board of Estimate acts for the Committee on Slum Clearance in directly supervising tenant relocation, management and maintenance activities of Title I sponsors. The Bureau maintains a supervisory office at each site and its staff carefully inspects all rental units in private housing to which Title I site families relocate. The field inspections and site occupant records maintained by the Bureau indicate conclusively that substantially all of the Title I site families were relocated to standard, decent, safe and sanitary dwellings, at rentals within their financial means.

These ten projects contained a total of 13,258 families at the time of acquisition. Of these, 10,645 families have been relocated as of July 1st, 1957. Public housing accommodated 2,237 or 21% of the families relocated, while 6,977 or 65.5% were relocated to private housing, mostly to rental units. Of this group 1,771 families or 16.6% of the total, were relocated to rental housing obtained by the redevelopers and 5,206 or 48.9% were relocated to private housing of their own choosing.

The Bureau of Real Estate, independently of the redeveloper, traces families vacating without leaving forwarding addresses and of the total number of families relocated from Title I sites 10% moved with whereabouts unknown and 3.6% relocated to substandard dwellings. This is reasonable for large scale relocation projects and compares favorably with the experience of the New York City Housing Authority. Of the 25,104 families relocated from City Housing Authority sites during the four year period ending September 1956, that Agency reported 11.3% vacated with whereabouts unknown and 7.5% as having relocated to sub-standard housing.

The Committee has recently established a practice of publishing reports to the public and to site tenants of Title I projects explaining relocation procedures and indicating by graphic charts the time schedules for relocation of families, demolition of existing structures and construction of the new development. The information contained in the reports should help site tenants to understand not only the benefits resulting from improvements but also the kind of assistance to be rendered.

SUMMARY

In addition to well planned, low-coverage, fireproof housing, the projects carry with them other important incidental improvements including schools, playgrounds, parks, new community facilities, shopping areas, and off-street parking and street widening to alleviate traffic congestion. The Columbus Circle project has provided the City with a much needed, centrally located Coliseum for shows, exhibits and conventions including a garage and parking facilities for 654 cars. Pratt Institute, Fort Greene and Washington Square Southeast will provide, in addition to new housing, areas for institutions of higher education and Fort Greene will also provide for expansion of private hospital facilities. The plans for N.Y.U.-Bellevue project include a professional building which will supply the needs of Bellevue Hospital and N.Y.U.-Medical School.

One of the largest and most widely discussed projects of the Title I program is the development of Lincoln Square. This project, estimated to cost about $150,000,000 covers an area of 45 acres in thirteen blocks, north and west of Columbus Circle and will include a "Housing Center," to provide apartment buildings for 4,500 families, a "Performing Arts Center," to include a new Metropolitan Opera House and houses for ballet and light opera, a new Philharmonic Symphony Concert Hall, the Juilliard Music School and related educational facilities, a "Collegiate Center" for the development of a new college and campus for Fordham University and a new public school, shopping centers and City parks. Included in projects being studied for future development, is Bellevue South, which will provide in addition to residential housing a needed Health Department laboratory to replace the obsolete and inadequate laboratory now at the site vacated by the Willard Parker Hospital, and the Cooper Square project which will provide for expansion of Cooper Union

PARK ROW

At top, Madison St., right, Park Row and bottom, St. James Place. Gov. Smith Houses to left, Municipal Building upper right.

Institute and an industrial development to meet the needs of light manufacturing now located in the area.

During the seven years since the urban redevelopment law was enacted by Congress, New York City has striven to take full advantage of assistance. Numerous obstacles had to be overcome to get the program off the ground, and there have been plenty of difficulties since then. Despite discouraging delays and obstructions encountered by project sponsors in their efforts to obtain F.H.A. commitments and financing, all operating projects, where financing was planned with F.H.A. guarantees, are now either under construction or ready to start. Other projects being constructed with conventional financing are moving along at a satisfactory pace. There are now encouraging indications that this long period of delay and adjustment is coming to an end with the establishment of workable Federal regulations and procedures which should result in much more rapid progress in the future.

Slum clearance under Title I is an important weapon in a bold attack on an old evil. It can be used to create adequate modern living communities within the City. The challenge is not only to rid ourselves of slums and to provide better housing facilities and public improvements by means of public and private enterprise, but to redesign the City in conformity with other major public plans and improvements as a series of pleasant communities for the families which still prefer to dwell in the City and enjoy the many advantages it offers.

CONCLUSION

The above record is disappointing, but by no means tragic or hopeless. It was probably too much to expect that exploitation and decay of a hundred or more years could be arrested, removed or reversed in a short time by the Title I program, involving as it does many public officials and public funds at several levels of government, vested interests of long standing, shrewd private sponsors, lenders and builders, thousands of tenants, the press and general public.

The recent delays are due to man failure, largely chargeable to Washington, but also brought about by the reluctance of local officials to move people in large numbers.

Certainly one of the conclusions we have arrived at as the result of our recent experience in this city is that it is unrealistic to imagine that the worst slums in the City, aggregating some five thousand acres, can be absolutely cleared and entirely new structures substituted in any forseeable time. The cost would be simply staggering, running into many billions of dollars. It is therefore necessary for all agencies in this field, including of course the City Planning Commission, to distinguish between absolute slum clearance, which is possible in some of the areas, and rehabilitation and improvement of conditions by one device or another in other less blighted areas.

The definition of the worst slums to be eliminated by clearance must be changed and limited. Let us assume that these can be reduced to two thousand acres. The remainder should then be set aside for improvement of conditions and rehabilitation by modernization, remodeling, repairing and enforcement of existing and new legislation. Responsibility for complete slum clearance should continue then to be divided between the State and City public housing agencies and those responsible for Title I and related private capital improvements. As to rehabilitation, and improvement of conditions, the problem gets down to raising standards by law and offering workable inducements to private owners and developers by tax reductions and other devices to compel or encourage them to undertake work which cannot possibly be carried out by public agencies. In this connection, it does not seem to us to be at all realistic to assume that public piecemeal redevelopment and rehabilitation of areas, including bits of new construction and large public undertaking of rehabilitation will in any foreseeable time accomplish very much in this field. The real job here will have to be done by raising standards in the Multiple Dwelling and other controlling laws and regulations and by offering substantial inducements to private owners and developers to go ahead on their own.

ROBERT MOSES, *Chairman*
THOMAS J. SHANAHAN, *Vice-Chairman*
JAMES FELT
PERCY GALE, JR.
BERNARD J. GILLROY
ROBERT G. MCCULLOUGH

SEWARD PARK

From Grand St. at left to East Broadway at right.
Seward Park below.

HAMMELS

At Rockaway Beach, Beach 90th St., at top, to Beach 74th St. Rockaway Beach Blvd. at right.

SEASIDE ROCKAWAY

At Rockaway Beach, Beach 102nd St. at top to Beach 108th St. Long Island Railroad at left.

BATTERY PARK

From Whitehall St., at left, to Coenties Slip, Water St. at rear. Ramps from South St. Viaduct and to Battery Park Underpass are in foreground. To left is Battery Park and to right, Jeanette Park.

SOUNDVIEW

Along East River in the Bronx. White Plains Rd. is at right and O'Brien St. above. The area being filled to left is extension of Soundview Park shown above and farther left.

MID-HARLEM

Between St. Nicholas Ave., to right, and Eighth Ave., from 135th St., below, to 125th St. Saint Nicholas Park is to right.

PENN STATION SOUTH

From 23rd St. at right to 29th St., from Ninth Ave. below to Eighth Ave. North point points at Pennsylvania Station.

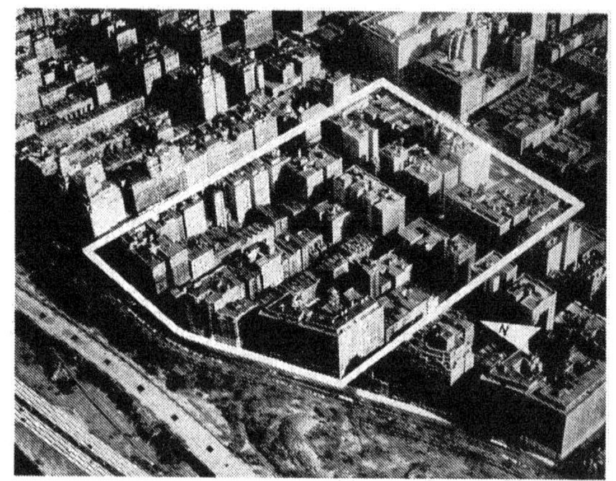

RIVERSIDE-AMSTERDAM

From 83rd St., at right, to 86th St. Amsterdam Ave. is in the background. Approximately one-half of the area is scheduled for slum clearance and is included in the project. Existing good buildings are to remain. In the foreground is Riverside Drive and Park and Henry Hudson Parkway.

JAMAICA RACE TRACK

In the foreground, left is Baisley Blvd., right, New York Boulevard. Long Island Railroad is to left. In background is 137th Ave.

CADMAN PLAZA

To the right is Henry St. and to the left, Fulton St. At top is Clark St. Cadman Plaza Park to the left includes the Red Cross Building and the Brooklyn War Memorial. Brooklyn Bridge approach stretches across the lower section of the picture.

COOPER SQUARE

To the left are the Bowery and Third Ave., to the right Chrystie St. and Second Ave. At top is 9th St. The southerly boundary, Delancey St., is just off picture below. The north arrow points at Cooper Union.

BELLEVUE SOUTH

The right area outlined below is bounded by Second Ave., to left, First Ave., to right, 23rd St., below and 30th St., above. Along East River and East River Drive, above right, are, from right to left, the Veterans Hospital, Bellevue Hospital and N.Y.U. Bellevue Medical Center.

GRAMERCY PARK

The left area outlined above is bounded by Third Ave., to left, Second Ave., to right, 24th St., below and 27th St., above.

HOUSING AND HOME FINANCE AGENCY OF THE UNITED STATES

ALBERT M. COLE, *Administrator*
RICHARD L. STEINER, *Commissioner, Urban Renewal Administration*

BOARD OF ESTIMATE OF THE CITY OF NEW YORK

ROBERT F. WAGNER, *Mayor*
LAWRENCE E. GEROSA, *Comptroller*
ABE STARK, *President, The Council*
HULAN E. JACK, *President, Borough of Manhattan*
JAMES J. LYONS, *President, Borough of The Bronx*
JOHN CASHMORE, *President, Borough of Brooklyn*
JAMES A. LUNDY, *President, Borough of Queens*
ALBERT V. MANISCALCO, *President, Borough of Richmond*

COMMITTEE ON SLUM CLEARANCE

ROBERT MOSES, *Chairman*
City Construction Co-Ordinator
THOMAS J. SHANAHAN, *Vice Chairman*
City Housing Authority
JAMES FELT
Chairman, City Planning Commission
PERCY GALE, JR.
Director of Real Estate — Board of Estimate
BERNARD J. GILLROY
Commissioner, Department of Buildings
ROBERT G. McCULLOUGH
Chief Engineer, Board of Estimate
GEORGE E. SPARGO,
Assistant to Chairman
WILLIAM S. LEBWOHL,
Director

CONSULTANTS

SKIDMORE, OWINGS & MERRILL, *Architects-Engineers, Coordinating Architects for all projects*
CHAPMAN, EVANS & DELEHANTY, *Architects*
EGGERS & HIGGINS, *Architects*
HARRISON & ABRAMOVITZ, *Architects*
W. K. HARRISON, H. H. GOLDSTONE, *Associated Architects*
KELLY & GRUZEN, *Architects*
S. J. KESSLER & SONS, *Architects and Engineers*
KAHN AND JACOBS, *Architects*
LEON AND LIONEL LEVY, *Architects*
LORIMER AND ROSE, *Architects-Engineers*
McKIM, MEAD & WHITE, *Architects*
HERMAN J. JESSOR, *Architects*
VOORHEES, WALKER, SMITH & SMITH, *Architects & Engineers*
WILLIAM WILSON, *Architect*, WM. I. HOHAUSER, *Associated Architect*
CHARLES F. NOYES COMPANY, INC., *Real Estate Consultants*
WOOD, DOLSON COMPANY, INC., *Real Estate Consultants*
FREDERICK E. MARX, *Real Estate Consultant*
MILTON SASLOW, *Tenant Relocation Consultant*
REAL ESTATE BOARD OF NEW YORK

SPONSORS — OPERATING PROJECTS

CORLEARS HOOK	EAST RIVER HOUSING CORP.
HARLEM	GODFREY NURSE HOUSES, INC.
NORTH HARLEM	HARLEM ESTATES, INC.
COLUMBUS CIRCLE	COLUMBUS CIRCLE APARTMENTS, INC.
	TRIBOROUGH BRIDGE AND TUNNEL AUTHORITY
MORNINGSIDE	MORNINGSIDE HEIGHTS HOUSING CORP.
WEST PARK	MANHATTANTOWN, INC.
FORT GREENE	THE BROOKLYN HOSPITAL
	LONG ISLAND UNIVERSITY
	KINGSVIEW HOMES, INC.
	UNIVERSITY TOWERS, INC.
PRATT INSTITUTE AREA	PRATT INSTITUTE
	HALL DEVELOPERS, INC.
N.Y.U.-BELLEVUE	UNIVERSITY CENTER, INC.
WASHINGTON SQUARE S.E.	NEW YORK UNIVERSITY
	WASHINGTON SQUARE VILLAGE CORP.

CITY OF NEW YORK — COMMITTEE ON SLUM CLEARANCE — TITLE I PROJECT — July 15, 1957

PROJECTS UNDER CONTRACT	ACQUISITION COST	WRITE DOWN & IMPROVEMENTS	FEDERAL SHARE	CITY SHARE	CONSTRUCTION COST	SITE IMPROVEMENTS
Corlears Hook	5,081,279	5,284,871	3,523,247	1,761,624	18,000,000	452,703
Harlem	5,244,383	7,404,494	4,855,059	2,549,435	19,000,000	2,549,435
North Harlem	4,801,513	5,586,836	3,805,828	1,781,008	17,000,000	1,217,736
West Park	15,396,682	16,699,729	11,133,153	5,566,576	34,000,000	3,509,683
Morningside	5,082,890	4,321,946	2,881,298	1,440,648	14,500,000	58,000
Columbus Circle	11,786,682	9,296,620	6,197,746	3,098,874	43,000,000**	55,000
Fort Greene	6,769,470	6,217,616	4,145,080	2,072,536	13,500,000**	544,454
Pratt Institute	7,700,123	8,183,062	5,455,375	2,727,687	26,000,000**	1,088,397
N.Y.U.-Bellevue	7,795,898	5,726,000	3,817,333	1,908,667	18,000,000	
Washington Sq. S.E.	19,286,481	15,416,532	10,277,688	5,138,844	42,000,000**	1,000,000
Sub Total	88,945,401	84,137,706	56,091,807	28,045,899	245,000,000	10,475,408

INVESTMENT, INCLUDING LAND & CONSTRUCTION BY PRIVATE CAPITAL $270,889,659
FEDERAL, CITY AND PRIVATE INVESTMENT 355,027,365

ADVANCED PLANNING	ACQUISITION COST	WRITE DOWN & IMPROVEMENTS	FEDERAL SHARE	CITY SHARE	CONSTRUCTION COST	SITE IMPROVEMENTS
Lincoln Square	40,000,000	42,185,000	28,123,334	14,061,666	150,000,000**	10,285,000
Seward Park	8,000,000	9,257,064	6,171,376	3,085,688	21,000,000	2,443,113
Park Row	3,432,000	4,722,226	3,148,150	1,574,076	7,200,000	936,933
Seaside	4,200,000	4,500,000	3,000,000	1,500,000	22,500,000	1,000,000
Hammels	7,000,000	6,000,000	4,000,000	2,000,000	31,000,000	1,500,000
Sub Total	62,632,000	66,664,290	44,442,860	22,221,430	231,700,000	16,165,046
PRELIMINARY PLANNING						
Penn Station South	18,000,000	18,000,000	12,000,000	6,000,000	31,000,000	1,000,000
Riverside-Amsterdam	9,000,000	8,220,000	5,480,000	2,740,000	20,000,000	220,000
Battery Park	7,700,000	6,900,000	4,600,000	2,300,000	12,000,000	100,000
Gramercy Park	6,200,000	6,042,000	4,028,000	2,014,000	18,000,000	750,000
Soundview	1,000,000	3,000,000	2,000,000	1,000,000	25,000,000	1,500,000
Jamaica Race Track	3,500,000	3,000,000	2,000,000	1,000,000	45,000,000	1,000,000
Sub Total	45,400,000	45,162,000	30,108,000	15,054,000	151,000,000	4,570,000
TOTAL	196,977,401	195,963,996	130,642,667	65,321,329	627,700,000	31,210,454

TOTAL INVESTMENT INCLUDING LAND & CONSTRUCTION BY PRIVATE CAPITAL $685,082,559
TOTAL FEDERAL, CITY AND PRIVATE INVESTMENT 881,046,555

UNDER STUDY	ACQUISITION COST	WRITE DOWN & IMPROVEMENTS	FEDERAL SHARE	CITY SHARE	CONSTRUCTION COST	SITE IMPROVEMENTS
Bellevue South	15,000,000	13,500,000	9,000,000	4,500,000	22,000,000	1,500,000
Cooper Square	22,000,000	21,000,000	14,000,000	7,000,000	50,000,000	1,500,000
Mid-Harlem	8,200,000	8,400,000	5,600,000	2,800,000	19,000,000	500,000
Cadman Plaza	5,500,000	5,500,000	3,700,000	1,800,000	12,000,000	500,000
Sub Total	50,700,000	48,400,000	32,300,000	16,100,000	103,000,000	4,000,000
FUTURE						
Seward Park Ext.	11,700,000	12,420,000	8,280,000	4,140,000	22,000,000	
Washington Market	17,542,000	15,000,000	10,000,000	5,000,000	27,000,000	
Park Row Ext.	2,012,000	2,148,000	1,432,000	716,000	4,000,000	
Division Street	2,700,000	4,200,000	2,800,000	1,400,000	7,500,000	
Arverne Rockaway	5,000,000	4,050,000	2,700,000	1,350,000	22,500,000	
Cathedral Parkway	6,600,000	6,000,000	4,000,000	2,000,000	14,000,000	
Mott Haven	2,236,000	2,250,000	1,500,000	750,000	12,000,000	
Northern Boulevard	1,500,000	1,500,000	1,000,000	500,000	27,000,000	
South Brooklyn	4,200,000	4,050,000	2,700,000	1,350,000	20,000,000	
Atlantic Avenue	1,500,000	1,500,000	1,000,000	500,000	13,000,000	
Sub Total	54,990,000	53,118,000	35,412,000	17,706,000	169,000,000	
TOTAL	105,690,000	101,518,000	67,712,000	33,806,000	272,000,000	

INVESTMENT, INCLUDING LAND & CONSTRUCTION BY PRIVATE CAPITAL $307,282,000
FEDERAL, CITY AND PRIVATE INVESTMENT 412,800,000

	ACQUISITION COST	WRITE DOWN & IMPROVEMENTS	FEDERAL SHARE	CITY SHARE	CONSTRUCTION COST	SITE IMPROVEMENTS
GRAND TOTAL	302,667,401	297,481,996	198,354,667	99,127,329	899,700,000	35,210,454

* Cooperative developments — includes tax abatement on all or part of project
** Includes Institutional development and Coliseum in Columbus Circle project

NET ACRES	DWELL'G UNITS EXIST.	NEW	TAXES OLD	NEW
14.66	878	1,668	104,640	104,640*
14.79	2,065	1,710	162,000	645,000
13.07	1,108	1,785	148,000	576,000
26.20	4,120	2,560	456,000	1,184,000
9.39	1,384	981	120,000	120,000*
6.28	286	608	305,800	346,000
21.20	510	836	145,000	278,000*
29.18	1,354	2,009	212,000	832,000
9.44	1,357	1,000	196,000	669,000
17.68	151	2,184	396,000	1,024,000
161.89	13,213	15,341	2,245,440	5,778,640
45.00	5,320	4,500	960,000	2,180,000*
15.21	1,494	1,704	222,000	437,000*
5.24	410	400	88,000	155,000*
35.00	142	1,500	111,000	768,000
49.22	1,763	2,107	185,000	1,075,000
149.67	9,129	10,211	1,566,000	4,615,000
21.78	2,250	2,400	486,000	680,000*
8.24	1,078	1,630	160,000	700,000
5.08	10	815	208,000	470,000
8.30	1,443	1,355	185,000	624,000
72.84	283	2,340	27,000	520,000*
178.00	0	4,000	127,000	950,000*
294.24	5,064	12,540	1,193,000	3,944,000
605.80	27,406	38,092	5,004,440	14,337,640
20.65	3,064	1,535	400,000	700,000*
27.52	2,277	1,800	595,000	1,400,000*
13.21	2,080	1,430	236,000	690,000
6.67	350	810	147,000	440,000
68.05	7,771	5,575	1,378,000	3,230,000
19.65	1,700	1,700	312,000	508,000*
17.50	26	1,800	468,000	1,080,000
2.66	207	250	54,000	92,000*
5.31	283	500	100,000	170,000*
34.64	32	1,500	132,000	800,000
10.00	1,600	1,000	176,000	528,000
8.40	315	800	60,000	400,000
24.00	0	1,800	47,000	560,000*
22.70	607	0	120,000	688,000
12.60	505	900	44,000	120,000*
157.46	5,275	10,250	1,513,000	4,946,000
225.51	13,046	15,825	2,891,000	8,176,000
831.31	40,452	53,917	7,895,440	22,513,640

TOTAL PROGRAM INVESTMENT, INCLUDING LAND AND CONSTRUCTION BY PRIVATE CAPITAL, $992,365,000

FEDERAL, CITY AND PRIVATE INVESTMENT $1,290,000,000

Richard C. Guthridge, Design

Aerial photographs by Skyviews, N. Y.

Printed in the USA
CPSIA information can be obtained
at www.ICGtesting.com
LVHW051904021123
762492LV00049B/409